Underwater Homes

THERESE HOPKINS

PowerKiDS press
New York

Published in 2009 by The Rosen Publishing Group, Inc.
29 East 21st Street, New York, NY 10010

Copyright © 2009 by The Rosen Publishing Group, Inc.

All rights reserved. No part of this book may be reproduced in any form without permission in writing from the publisher, except by a reviewer.

First Edition

Editor: Nicole Pristash
Book Design: Kate Laczynski
Photo Researcher: Jessica Gerweck

Photo Credits: Cover, pp. 1 © Brandon Cole/Getty Images; p. 5 © www.istockphoto.com/Marco Crisari; pp. 7, 9 © Norbert Wu/Getty Images; p. 11 © www.istockphoto.com/John Anderson; pp. 13, 15, 17, 19, 23, 24 Shutterstock.com; p. 21 © Georgette Douwma/Getty Images.

Library of Congress Cataloging-in-Publication Data

Hopkins, Therese.
　Underwater homes / Therese Hopkins. — 1st ed.
　　　p. cm. — (Home sweet home)
　Includes index.
　ISBN 978-1-4358-2694-6 (library binding) — ISBN 978-1-4358-3068-4 (pbk.)
　ISBN 978-1-4358-3080-6 (6-pack)
　1. Marine animals—Habitat—Juvenile literature. I. Title.
　QL122.2.H36678 2009
　591.77—dc22
　　　　　　　　　　　　　　　　　2008020794

Manufactured in the United States of America

CONTENTS

Underwater Homes 4

Sea Grasses 6

Coral Reefs 10

Kelp Forests 18

Words to Know 24

Index 24

Web Sites 24

Many animals live underwater. This sea star makes its home in sea grasses.

Sea grasses look like the grass you may find in a **meadow** or in your backyard.

Yellow stingrays swim through sea grasses to find sea worms and crabs to eat.

A **coral reef** is an underwater home where many fish, **sponges**, and snails live.

Coral comes in many different shapes and colors.

This sea turtle is swimming through a coral reef looking for its lunch.

Parrot fish eat the coral itself.

Fish and other sea animals hide and feed among the **kelp** leaves in this kelp forest.

Sea urchins live in kelp forests, but urchins also like to eat the kelp.

Sea otters like kelp because they eat many of the animals that live in kelp forests.

WORDS TO KNOW

coral reef

kelp

meadow

sponges

INDEX

WEB SITES

A
animals, 4, 18, 22

C
coral, 12, 16
coral reef, 10, 14

F
fish, 10, 16, 18

K
kelp, 20, 22
kelp forest(s), 18, 20, 22

Due to the changing nature of Internet links, PowerKids Press has developed an online list of Web sites related to the subject of this book. This site is updated regularly. Please use this link to access the list:
www.powerkidslinks.com/hsh/uwater/

24